Alfred Edward Mathews

Gems of Rocky Mountain scenery

Alfred Edward Mathews

Gems of Rocky Mountain scenery

ISBN/EAN: 9783337717896

Printed in Europe, USA, Canada, Australia, Japan

Cover: Foto ©ninafisch / pixelio.de

More available books at **www.hansebooks.com**

GEMS

OF

ROCKY MOUNTAIN

SCENERY,

CONTAINING VIEWS ALONG AND NEAR THE

UNION PACIFIC RAILROAD;

BY

ALFRED E. MATHEWS.

New-York:
PUBLISHED BY THE AUTHOR,
1227 Broadway.

1869.

Entered according to Act of Congress, in the year 1859, by A. E. Mathews, in the Clerk's office of the District Court of the United States, for the Southern District of New-York.

THE NEWBERRY
LIBRARY

PRESS OF J. ADNAH SACKETT, 60 JOHN STREET, NEW-YORK.

INTRODUCTORY.

THE Lithographs embodied in this work are selections from a series of sketches made by the artist while sojourning in Colorado, Idaho, Montana and Utah, from the fall of 1865 to the winter of 1868. During this time he made many excursions of more or less duration, from Denver in Colorado, Helena and Virginia City in Montana, and Salt Lake City in Utah; the entire distance accomplished being about 6,000 miles; remaining, however, but one winter in the mountains. These expeditions were performed, excepting during one summer, entirely alone, and principally with ponies; but on two or three occasions on snow-shoes and in a small boat. One pony was used for riding—the other carried a small, light tent, bedding and provisions. Equipped in this way the artist was prepared to camp wherever and whenever so inclined—the tent being a perfect security against wild animals at night.

The pictures represent actual localities; and as they have been drawn on stone from the sketches by the artist himself, have lost none of their original truthfulness.

It will be observed that quite a large number of the scenes represented are located in Colorado; this is because a larger proportion of the sublime and beautiful mountain scenery of the great Rocky Mountain belt cluster together in this incomparable State. The Territories represented are arranged in alphabetical order.

It would require many, very many, volumes to represent the half of the numerous, grand and awe-inspiring views that are scattered so profusely throughout the entire length of this vast belt of mountains; so that an apology for leaving out some justly celebrated and comparatively well known localities is, perhaps, scarcely necessary.

ALFRED E. MATHEWS.

COLORADO.

THE EASTERN SLOPE.

THIS picture represents a small section only of the eastern slope of the Rocky Mountains near Denver, the City of the Plains. The sketch was taken on the Great American Plains, about 14 miles from Denver.

This portion of the eastern slope, above all others which it has been the privilege of the artist to behold, is peculiar for the massiveness of its rocks that project up from the mountain sides at all angles, forming strong and pleasing contrasts of light and shade, that are at times truly sublime. The strength of outline of these mountains, and the strong contrasts of light and shade produced by the massiveness and form of the gigantic rocks, gives them additional beauty and intensity of color, that varies according to the condition of the atmosphere and the position of the sun and clouds.

The two, or rather three peaks seen on the left, constitute one mountain, called Saddle Mountain, from its supposed resemblance (when seen from the Denver and Golden City road) to a saddle. Distance from the point of observation to Saddle Mountain six miles. On the right, fifty miles or more distant, the celebrated Long's Peak pierces the clouds.

BEAR CANYON,

IS grand and awe inspiring beyond description, and in passing up beneath its massive and towering walls, one feels as if he was walking beneath the arched dome of some vast cathedral, or in the presence of the Creator, and instinctively uncovers his head. The crevices of the rocks are decorated with pines, and the deep gorges are made still darker by their dense foliage—fitting haunts of bears and mountain lions, that skulk through the ravines. The view is from the rocky side of one of the foot hills that are here unusually large, rugged and numerous, immediately opposite the mouth of the canyon. A small stream rushes wildly down the canyon over a rough bottom, and an old deserted road winds up to the distant mountain-tops. This road is sometimes used to convey logs down to the saw-mill a short distance below. Bear Canyon is distant about three miles from Boulder, and twenty-one miles from Denver.

In the fall of 1868, a young man was here torn to pieces and devoured by mountain lions. This animal is a kind of cougar or puma, and is nearly the size of a full grown lioness.

An immense rock is seen near the road on the left of the picture,—under this the artist pitched his tent while in the locality.

THE SIERRA MADRE.

DOUBTLESS, the finest view of the Great Sierra Madre is to be had near Columbia, in Ward District. The scene represented is from Red Rock Lake, about four miles from Columbia, and represents the Range as seen in the month of July, 1867. The large mountain on the right is St. James' Peak.

This is the main range of the Rocky Mountains, which runs through the centre of Colorado in a serpentine course, sending out numerous spurs, and enclosing four large and beautiful parks: the North, Middle, South, and San Luis Parks. West of this chain of parks, enclosed with a granite wall, the country is most beautifully diversified with mountain ranges, broad valleys, elevated plateaus, lakes, rivers, parks and wooded slopes, but is comparatively little known; and except on ponies or mules, difficult of access, and is the hunting ground of the friendly Ute Indians.

One of the most beautiful features of the Snowy Range is the numerous flowers that are often seen entirely surrounded by, and almost covered with snow. Many of these flowers are of most peculiar and exquisite beauty.

CLEAR CREEK CANYON.

THIS picture is a view in Clear Creek Canyon, at a point five miles above Golden City, looking down, and is one of the many truly magnificent sights that have rendered Clear Creek Canyon so justly celebrated for its sublimity of scenery.

However grand and imposing in appearance as a whole, it is very seldom that views in canyons, when transferred to paper, convey a pleasing effect; this is owing to the difficulty of presenting a picturesque and at the same time truthful arrangement of light and shade; but in this view, it will be seen, there is a happy combination of light, shade and form, giving to the view a most beautiful effect. This effect is seen about ten o'clock in the morning.

THE CHIEF, SQUAW AND PAPOOSE.

THESE are among the most interesting mountains in Central Colorado, east of the "Range," and are situated from five to ten miles north of Idaho. In the lithograph, the Chief (seen on the right) is the most distant mountain, and is higher than the others. The Squaw is represented on the left, while between the two is the Papoose. The view is from the head of Virginia canyon, three miles from Central City. Part of the mountain slopes that enclose the upper extremity of Virginia canyon can be seen on either side in the middle distance of the picture.

CHICAGO LAKES.

HERE is represented one of the grandest and most beautiful localities in Colorado. Two small but lovely lakes, almost surrounded by high rocky walls, nestle among the peaks, near the summit of the Great Snowy Range. Their waters are fed and kept icy cold by the drippings from immense banks of snow and ice that are scattered over the surrounding mountains, and hidden in deep gorges. They are on the verge of timber line, and the upper lake is elevated above the lower one, so that from the point of observation it was out of sight. The sunny exposures near them are in midsummer covered with rich grass, beautiful mosses, and lovely flowers. The lakes have been named in honor of the City of Chicago.

The water of these two Lakes is so clear that the trout, quite numerous in the lower one, can be seen gliding about with almost as much distinctness as birds can be seen flying through the air.

It is impossible for the human eye to appreciate the immensity and altitude of the Great Rocky Range; it is only by climbing their rugged slopes that we can arrive at a true estimate of their vastness. In visiting Chicago Lakes the traveler follows up first Clear Creek and then Chicago Creek for *sixty miles* from the base of the mountains, and for the entire distance the water of these two creeks is white with foam in tumbling over rocks and mountain gorges on its way to the great plains—he is then only at the foot of the mighty range itself.

THE OLD MOUNTAINEER,

IS on South Clear Creek, immediately opposite the mouth of Fall River, two and a half miles above Idaho, and stands 234 feet in perpendicular height above the level of Clear Creek. The profile is eighteen feet in length from the chin to the crown of the head. Openings between the rocks form the eye and mouth, and a cluster of mountain bushes on the top of the head represents the hair. This profile is a more complete representation of the "human face divine" than the celebrated "Old Man of the Mountains" of the White Mountains of New Hampshire, although it is not in so conspicuous a position. It varies in shape and expression according to the point from which it is seen, but the best view can be had close to the water, on the northern shore of Clear Creek, a few rods above Fall River Bridge.

This natural curiosity was discovered by the artist in 1865. The elements cannot fail to have some effect on rocks, however hard, that are so exposed, and the features are liable at any time to be changed in their resemblance to the human face, or entirely destroyed.

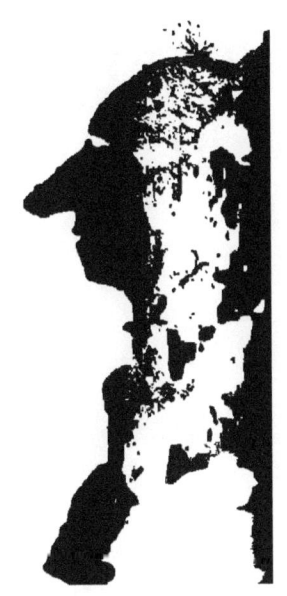

GRAY'S PEAK.

THIS is supposed to be the second highest mountain in Montana, its approximate altitude being 14,700 feet above tide water. It branches off towards the summit into two peaks; the highest of the two is seen back and to the right of the one prominently shown. The view is from near the summit of Snake River Pass, and represents the mountain as seen in the month of July, 1867. The cold was hightened to such intensity by the wind that it was impossible to remain while sketching more than a minute or two at a time, without running over to the sunny side of the pass to get warmed in the sun's rays. Rich silver mines cover the sides of this and the adjoining peaks.

Two thousand feet below the spot shown in the foreground of the picture—far down in the deep gorge on the left—the artist crossed a small foaming torrent with ponies over a bridge of pure snow : this was in the month of July. An unusual quantity of snow had fallen during the winter and spring preceding.

A most magnificent view is had from the summit of the Snake River Pass : to the north and northwest, as far as the eye can reach, is a perfect ocean of rugged, snow-clad peaks, full of deep gorges and rock-bound lakes ; awful in their grandeur, with everywhere the most complete stillness and solitude. Towards the east the character of the view is entirely different : equally grand, but less sublime—vaster, but less imposing. The prevailing colors are warmer, more beautiful, cheerful and soothing. Far in the back ground the plains look like a vast ocean of silvery blue ; while the mountains that roll down from the great Rocky Range are, first gray, then purple, then blue, with all degrees of light and shade, and strength of color, with varying form and outline.

Gray's Peak lies near the center of Colorado, in the very heart of the rich silver mines, fifteen miles from Georgetown.

BUFFALO MOUNTAIN.

THIS picturesque mountain is west of the great range, and is part of the western wall of the Blue River valley, near where Ten Mile Creek joins that river. Buffalo Mountain is the one towards the left in the picture. The sketch was taken near Elk Creek, a short distance below the mouth of Snake River, in Summit County.

At certain times, and with peculiar effects of light and shade, this scene possesses more desirable features for a picture than any other locality it has ever yet been the pleasure of the artist to visit.

In the mountains of this region are mountain sheep—the most delicious of all wild game,—and bears. In the valleys, antelope and elk; while the streams are full of brook trout.

TURKEY CREEK CANYON.

THE beauty of the numerous scenes which meet the eye of the traveler on his way from Denver to the South Park, is enhanced by an excellent road, recently built, through Turkey Creek Canyon.

The view represented is near the mouth of the canyon, looking up, or more properly across, and is about sixteen miles from Denver. The mountains in this part of Colorado contain numerous small and beautiful parks, and elevated valleys covered with the richest grass, beautifully intersperced with belts, patches and picturesque groups of pine trees; and the ravines and canyons are watered by clear and limpid streams.

EXIT OF THE SOUTH PLATTE FROM THE MOUNTAINS.

THE sketch here represented was taken from the low ridge of hills that extend along at the base of the mountains, parallel with them, commonly called the "Hog Back." This ridge extends in some places for miles near the mountains, forming beautiful, grassy valleys, half a mile or more in width, with occasional gaps, more or less wide, that allow of the exit of brooks, creeks and rivers from the mountains. Sometimes this ridge sinks beneath the surface of the earth to again reappear.

The South Platte takes its rise in the deep gorges of the snow-clad peaks that surround the South Park. Its tributaries are numerous, clear, and filled with trout. South and southeast of this park cluster groups of elevated valleys, thickly carpeted with grass, flowers and herbs; while the mountain slopes are generally well wooded with pine and fir trees of a small growth.

Distance of the locality represented in the picture from Denver is nineteen miles. The volume of water here is considerable, and rushes from the mountains with terrific force.

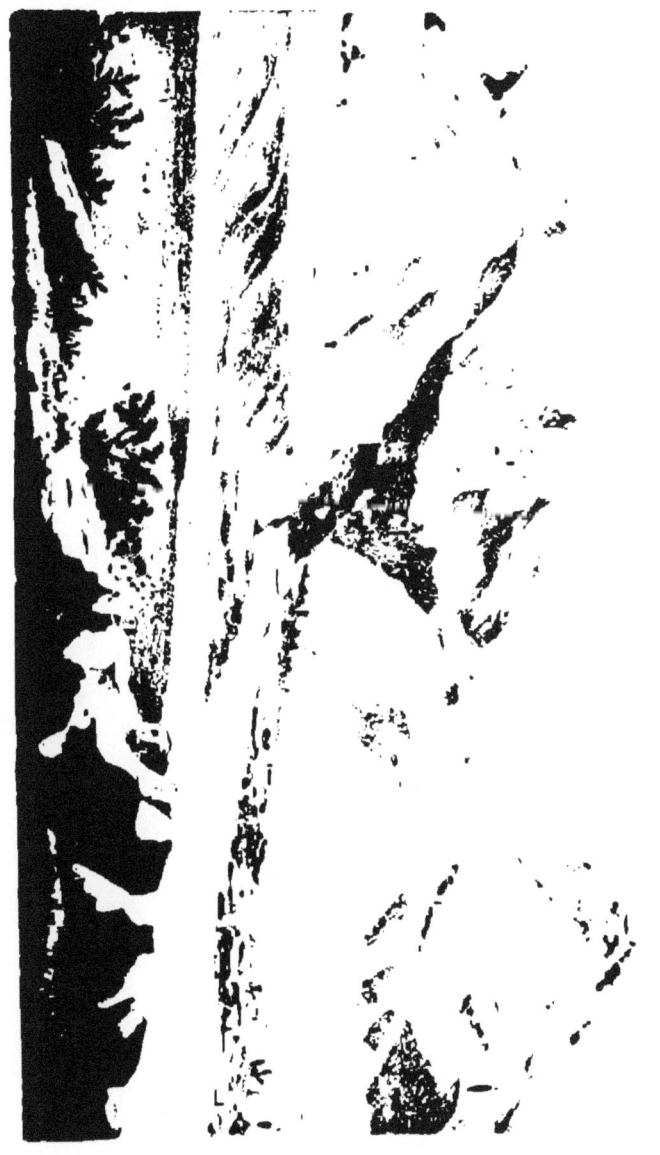

NATURAL MONUMENTS.

THE curious and beautiful in nature are here combined in a most pleasing manner. These monuments are near Monument Creek, which runs along parallel with the eastern slope of the mountains, about eight miles from them. They are to be seen on the slopes of the ridges which run down from the mountains; and they spread over an area of country several miles in extent. The Monuments represented are distant eight miles from Colorado City, five miles from the mountains, and sixty-eight miles south of Denver.

These Monuments are formed by the elements, and consist of a conglomerate of gravel, coarse sand, clay, and a little lime. The surface of the earth was at one time even with the top of the monuments, and in places was hardened by being exposed to the sun. The hard spots served as a protection to the material beneath, while the action of the rain gradually wore the rest away. Softer streaks of this conglomerate running in a horizontal direction, caused them to assume a variety of curious shapes. Some look like sibley tents; and those that show themselves between the trees suggest a vast encampment of soldiers in the groves of pine; others look like ladies with broad hats. Then there are representations of vases, urns, bells, bottles and pyramids; some of the latter inverted. The singular process by which nature forms them is still going on, and can be studied by those so disposed.

IDAHO.

A MIRAGE ON THE PLAINS.

THE traveler on the journey from Ogden to Montana Territory, if he will watch the western horizon, while crossing the Great Lava Plains of Snake River, is almost sure to witness one or more of these optical delusions of nature. They are of great variety, and occur principally in the morning, soon after or about sunrise, and late in the afternoon.

The most common mirage seen on these plains is the changing of form of some objects in sight along and near the horizon, and the elongation or looming up of others. The mirage represented in the plate is the Three Buttes, changed in form—two of them representing huge buildings with perfectly straight walls, towers and regularly sloping roofs; one of them having a singular shaped cap on the comb of the roof.

The cause which produces this phenomenon is a considerable difference of temperature in the strata of air, giving them very unequal densities; thus raising up objects or parts of objects, either perpendicularly or at various angles, according to the density of the stratum.

Sometimes immense fields of sage brush are made to loom up so as to assume in the eyes of the beholder the dignity and appearance of forests of cedar; and at other times, beautiful lakes are produced, filled with islands and surrounded by mountains that are taller than those commonly to be seen.

The Snake River Lava Plains are of great extent, and for the most part sandy, dry, and covered with sage brush. In places the vast bed of lava comes to the surface, extending over many miles, and is often filled with caverns, that are the haunts of wolves and other wild animals.

The Three Buttes are small mountains with very steep slopes lying to the west of the Montana road.

THE THREE TETONS.

THESE singular mountains are justly ranked among the high peaks of the complicated ranges that form the Rocky Mountain Belt, and are located in the Territory of Idaho. They are exceedingly rugged, and abound in frightful precipices with corresponding chasms. The view is from the Salt Lake and Montana stage road, near Kamas Creek, in Idaho Territory, looking east, and represents these peaks at a distance of about 125 miles. Their slopes are so steep that snow does not remain on the exposed parts long, but is soon blown off by the wind into their numerous and deep gorges. The name has been given them by the Shoshone Indians, and signifies "Woman's Breasts."

MONTANA.

Exit of the Yellowstone from the Mountains.

NOTHING can exceed the grandeur of the mountain scenery of the upper Yellowstone. The view represents its exit from these mountains, as seen from a point three miles below, and thirty miles from Bozeman, in the Territory of Montana. At the time the sketch was made (1867) no white inhabitants lived in the valley below the canyon, but several mining camps had been established in the mountains along the river and some of its tributaries. In the foreground of the view two antelopes are seen. These animals are quite numerous in the region, and during two days that the artist remained in the valley, he saw many large and small herds. Elk and mountain sheep also abound, and have frequently been seen in immense droves.

The Yellowstone is one of the most peculiar rivers on the continent, and is 1,600 miles in length. It is this stream and some of its tributaries that give to the Missouri its turbid appearance. Their waters, however, are all clear until nearing the Bad Lands—a region destitute of vegetation, and without springs or small streams.

This barren waste is thickly strewn with animal and vegetable petrifactions, and curious stones; and has been little explored. It is very similar in some respects to the country along Bitter Creek and Duck Lake on the Denver and Salt Lake road.

The source of the Yellowstone is a clear, deep, beautiful lake, far up among the clouds; that is kept cool by drippings from the eternal glaciers. Near this lake the river makes a tremendous leap down a perpendicular wall of rock, forming one of the highest and most magnificent waterfalls in America.

CITADEL ROCK,

ALSO in Montana, is the most prominent and one of the most singular of the many curious formations of the upper Missouri. It projects slightly out into the stream, and is of volcanic origin, containing argillaceous nodules and quartz crystals. The view is from the opposite bank, looking down. Near the foreground of the picture some elk are seen. Elk, antelope and deer frequently come down to the river to drink; and an animal that appears to be a kind of cross between a wild sheep and goat inhabits the bluffs. Bears skulk in willow thickets in the bottoms and ravines, and gray wolves and coyotes are numerous. The Bottom Lands are generally alluvial, and confined to small patches, or narrow strips along the river, that are covered with rank grass with here and there a scanty growth of cottonwood and willow trees. Beaver are very numerous along the banks and have cut down much of the timber.

The course of the Missouri is very crooked, running to and from all points of the compass, and is shut in by high bluffs, principally of sandstone, in many places perpendicular, out of which the elements are constantly carving pyramids, monuments, castles, fortresses, churches, and other singular representations. Beyond, rise ranges of naked hills, stretching out into broad, rolling plains, covered with short, nutritious grass.

In floating noiselessly down with the current in a small skiff, as the artist did, the shifting gravel and sand on the bed of the river made a peculiar grating sound. The mountains and plains of the northwest are gradually but constantly being worn away and carried down to build up a continent in the south.

UTAH.

CHURCH BUTTES.

THIS truly singular formation is near the old overland stage road, within eight miles of the Union Pacific Railway, and forty-five miles from Green River, in the Territory of Utah. The sketch was made when Wells, Fargo & Co.'s coaches were running, and represents one of them going west.

The material of Church Buttes is decomposed sandstone, of an ash color, with occasional strata of a slightly warm color, that has been thus formed by a gradual wearing and washing away of the surrounding surface. This region of country contains many similar bluffs and buttes. It is here that the beautiful and valuable moss-agates are found, although they are now becoming scarce.

ECHO CANYON,

UTAH Territory, has been so called from the fact that in some parts of it most perfect echos are produced. Echo Creek, that runs down the canyon, is a tributary of Weber River.

Next to Weber Canyon the scenery is perhaps finer than anywhere else between the Salt Lake Valley and Chyenne. The view represented is about three miles from the mouth of the canyon, looking down, and forty-five miles from Ogden. The Union Pacific Railroad passes through the valley, and part of a train of cars is represented in sight on the left, going west.

This canyon is twenty-five miles long, and the most singular feature about it is, that the perpendicular walls are confined entirely to the north side.

The mountains in this region are comparatively small, and are part of the vast ocean of spurs and foothills of the main range on the one side, and the Wasatch Range on the other. Their surface is rugged, dry, and comparatively barren; the only profuse growth being sage brush. Grass is scarce and short, and the few trees that struggle for a precarious existence are mostly hidden from sight in the bottom of canyons.

WEBER CANYON,
Looking Down.

WEBER River runs through a deep gorge that cuts across the Wasatch Range, and flows into Great Salt Lake, near Ogden.

The Wasatch Range is the eastern wall that encloses the Great Salt Lake Valley, and, as seen from the valley, is very picturesque, rising as it does, with remarkably steep slopes, in many places to the altitude of 4,000 feet above the lake—8,000 feet above tide water.

The mountains of this range are exceedingly rugged, and filled with deep gorges and canyons, with high, precipitous walls; and the wind that often rushes down them with great violence, is chilled by passing over the immense banks of snow always remaining on the higher peaks during summer.

The view here shown is from the Devil's Gate, looking down, and shows the railroad bridge that crosses the river just below. The approximate altitude of the most distant mountain on the right is 3,000 feet above the river.

The comparatively few trees that grow on these mountains are mostly hidden from view in the many deep, dark chasms.

WEBER CANYON,
Looking Up.

THIS view of Weber Canyon is had from near the railroad bridge shown in the preceding picture, just below the Devil's Gate. The old overland stage road winds along the canyon near the river.

The high rocky walls of this canyon are too massive in form—too great in altitude—too vast in extent and proportion to be adequately represented with the pencil, and too overpowering in their grandeur to be justly described with the pen. Weber River Canyon ranks first among the many fine views to be met with along the line of the Union Pacific Railroad.

APPENDIX.

Some of the former productions of Mr. Mathews have been endorsed by many of our leading men, among which are the following:

President GRANT says: "They are the most accurate and true to life I have ever seen."

Major-General GEORGE H. THOMAS says: "They are very accurate and life-like."

Major-General JAMES B. McPHERSON says: "I most cheerfully bear testimony to their general fidelity and spirit."

In reference to pictures of the Rocky Mountains by the same artist, the

Hon. G. M. DODGE says: "They are very accurate and artistically gotten up."

Colonel McCLURE says: "They are alike chosen and executed with great skill and fidelity."

The Rocky Mountain News says: "Mr. Mathews is the only artist that has yet done justice to the splendid scenery of our Territory."

www.ingramcontent.com/pod-product-compliance
Lightning Source LLC
Chambersburg PA
CBHW031605110426
42742CB00037B/1220